1

I. Introduction

The empirical literature on the competitive effects of consummated petroleum mergers is limited and its conclusions are mixed.[1] The appropriate quantitative methodology to identify competitive effects is also debated among analysts.[2] More retrospective of petroleum mergers are clearly warranted in view of continued widespread interest in the competitiveness of the industry. The federal agency in charge of antitrust merger enforcement involving petroleum, the Federal Trade Commission ("FTC") also recognizes that merger retrospectives serve to inform its own antitrust enforcement decisions (US FTC (2004)). The U.S. Government Accountability Office ("GAO") has similarly concluded that more retrospectives could enhance the FTC's mission to maintain competition in the petroleum (GAO (2008)). Retrospective studies may be especially informative on the appropriateness of key presumptions and conclusions of earlier, prospective antitrust analyses of the same transactions (Carlton (2009)).

This paper examines Sunoco's 2004 acquisition of El Paso's Eagle Point, New Jersey refinery, and Valero's 2005 acquisition of Premcor's refinery in Wilmington, Delaware. We examine the FTC's enforcement rationale in these mergers and test for post-merger price changes. Unlike previous petroleum merger retrospectives, which were restricted to gasoline, we test also for competitive effects in diesel fuel.

---

[1] See U.S. Government Accounting Office (2004), Hastings (2004), Taylor and Hosken (2007), Simpson and Taylor (2008), and Taylor, Kreisle, and Zimmerman (forthcoming).

[2] See U.S. Federal Trade Commission, Papers and Proceedings of Federal Trade Commission Conference, *"Estimating The Price Effects and Concentration in the Petroleum Industry: Evaluation of Recent Learning,* available at http://www.ftc.gov/ftc/workshops/oilmergers/index.shtm. For a more general overview of merger retrospectives, including those involving other industries, see Hunter, Leonard, and Olley (2008).

Section II provides background on the transactions and on Northeast bulk supply conditions in gasoline and diesel. Section III identifies the key competitive issues raised by the transactions. Section IV describes our methodology and data, while Section V presents our empirical findings. Section VI summarizes.

II. Background.

A. The Transactions

1. Sunoco/El Paso. Sunoco, Inc. ("Sunoco") bought El Paso's Eagle Point, New Jersey refinery on January 13, 2004 for about $250 million.[3] At the time, Sunoco operated refineries in Marcus Hook, PA, Philadelphia, PA, Toledo, OH and Tulsa, OK. The firm's logistical unit, Sunoco Logistical Partners, L.P. operated refined product and crude oil pipelines and terminals, primarily in the Northeast, Midwest and South Central parts of the U.S. A major branded retail marketer of gasoline and diesel fuel, Sunoco had 4,528 retail outlets in 25 states at the end of 2003. These retail outlets were concentrated in Connecticut, Massachusetts, New, York, New Jersey, Pennsylvania, Ohio, Michigan, and Florida. Sunoco refineries produced more than its branded retail operations sold. For example, in 2003 Sunoco's four refineries produced an average of 376 thousand barrels per day ("MBD") of gasoline and had sales to unaffiliated customers of 153 MBD.[4]

Primarily concerned with natural gas and electricity, El Paso entered petroleum refining in 2001 by acquiring the Coastal Corporation. Coastal owned four petroleum

---

[3] The transaction included the purchase of inventory and related assets. Related assets included ship and barge docks, product truck racks, and a 4.5 mile product pipeline from the Eagle Point refinery to the Harbor pipeline, which runs north to Linden, New Jersey. Sunoco, Inc. sold these related assets to Sunoco Logistics Partners L.P for $20 million in March 2004. Sunoco Logistics was a Master Limited Partnership formed by Sunoco, Inc. to own and operate petroleum pipelines, terminals and storage facilities. Sunoco Logistics acquired El Paso's 33.3 percent share in the Harbor Pipeline for $7.3 million in June 2004. See Sunoco Logistics News Releases, March 30, 2004 and June 29, 2004.

[4] Sunoco, Inc. 2003 10-K, March 5, 2004 at 2-7.

refineries and related marketing assets, as well as assets in the natural gas and chemical industries. Not long after buying Coastal, El Paso began to sell Coastal assets, and in 2003 the firm announced its intention to exit from the petroleum industry.[5] The Eagle Point sale reflected that decision. At the time of the sale to Sunoco, Eagle Point was purely a "merchant" refiner, selling all its output to unaffiliated customers. The Eagle Point refinery was about 15 miles away from Sunoco's two refineries in Philadelphia.

2. Valero/Premcor. Valero Energy Corporation ("Valero") acquired Premcor on September 1, 2005 for approximately $6.9 billion.[6] Valero had 17 refineries at the time of the Premcor transaction, including a refinery and an associated product terminal in Paulsboro, New Jersey. Valero also transported and stored petroleum products in many parts of the U.S. Its refineries supplied both unaffiliated customers and Valero branded distributors and retail dealers. At the time of the merger, Valero controlled about 3,000 retail sites nationwide. Valero also sold gasoline and diesel through company owned and operated branded retail outlets nationwide. There were approximately 1000 Valero-branded outlets at the time of the Premcor merger.[7]

Premcor had four refineries and was a leading seller of unbranded refined petroleum products. Its refineries were located in Port Arthur, Texas, Memphis, Tennessee, Lima, Ohio, and Wilmington, Delaware. Premcor had no branded wholesale or retail operations when it was bought by Valero. Premcor acquired the Wilmington

---

[5] El Paso sold Coastal's refineries in Aruba and Corpus Christi, Texas to Valero in 2002 and 2004 respectively. El Paso sold Coastal's Mobile, Alabama refinery to Trigeant in 2003. See El Paso Corporation 10-K Filings, March 31, 2003 and September 30, 2004.
[6] Valero Energy Corporation, 2005 10-K, March 1, 2006 at 2.
[7] Valero Energy Corporation, 2005 10-K, March 1, 2006 at 3-11.

refinery from Motiva on May 1, 2004.[8] Premcor's Wilmington refinery was a merchant

refinery, selling all of its output to unaffiliated customers, and was located approximately

35 miles downstream on the Delaware River from Valero's Paulsboro refinery.

Nearly coincident with Valero/Premcor, Valero proposed to buy Kaneb Pipeline

Partners, L.P., and a related entity, Kaneb Services LLC (hereafter collectively referred to

as "Kaneb") in November 2004.[9] Having operations in many parts of the U.S., Kaneb

was engaged in transporting and terminalling crude oil, intermediate and finished refinery

products. Kaneb sold no gasoline or other refined products in the Northeast and derived

revenue exclusively from transportation and terminalling services sold to refiners and

other marketers of refined products.[10] The combination of Kaneb's three Philadelphia

area product terminals and Valero's nearby refinery and product terminal was challenged

by the FTC.[11] The FTC alleged the proposed merger would result in anticompetitive

effects in both the sale of gasoline and in terminalling services in the greater Philadelphia

---

[8] Motiva was a joint venture between Royal Dutch Shell and Saudi Aramco. Premcor did not acquire any marketing assets from Motiva—which included the venture's Shell brand wholesale and retail operations-- as part of the Wilmington transaction. See Premcor Inc., 2004 10-K at 12.

[9] The actual transaction was proposed by Valero Limited Partners ("Valero L.P."), a master limited partnership formed by Valero Energy Corporation. Valero L.P was engaged in the transportation and storage of crude oil and refined petroleum products, and at the time derived about 98% of its total revenues from Valero Energy, which owned 46% of Valero L.P.'s common units. When analyzing the transaction, the FTC treated these Valero entities as one. Under the agreement, Valero L.P. would pay $525 million in cash and exchange $1.7 billion in Valero L.P. Partnership units for the Kaneb Pipeline Partners partnership units. Kaneb Pipeline Partners was a publicly traded limited partnership that owned Kaneb's product pipeline and petroleum product and specialty liquids storage and terminal facilities, including its wholly owned subsidiary, Kaneb Pipeline Company which operated these assets. Kaneb Services LLC owned the general partnership in Kaneb Pipeline Partners, as well as 5 million shares of Kaneb Pipeline Partner's limited partnership units. For purposes of its competitive analysis the FTC treated these related Kaneb entities as one. See Analysis of Proposed Consent Order to Aid Public Comment, available at: http://www.ftc.gov/os/caselist/0510022/050615anal0510022.pdf.

[10] See Complaint in the Matter of Valero L.P et al., Docket No. C-4141 (June 14, 2005) (hereafter referred to as the FTC Valero/Kaneb complaint) at paragraph 31, available at, http://www.ftc.gov/os/caselist/0510022/050615comp0510022.pdf.
Terminalling services involve the storage and throughput of petroleum products, including the dispensing of product to tank trucks for delivery to retail outlets.

[11] The FTC also challenged Valero's acquisition of Kaneb assets in the Colorado Front Range and in Northern California. See FTC Press Release, June 15, 2005, available at: http://www.ftc.gov/opa/2005/06/valerokaneb.shtm.

area.[12] The FTC entered into a consent agreement with Valero that required divestiture of Kaneb's North and South Philadelphia terminals and Kaneb's Paulsboro, New Jersey terminal. Pacific Energy Partners bought these assets on September 30, 2005. Though this divestiture maintained the pre-merger status quo in terminals and therefore anticompetitive price effects might not be expected, the FTC's challenge of the Valero/Kaneb merger provides additional insight on the agency's enforcement rationale in Sunoco/El Paso and Valero/Premcor, as we discuss below.

B. *Bulk Supply in the U.S. Northeast.*

The mergers' possible competitive implications can be better appreciated with an overview of bulk supply conditions in the U.S. Northeast at the time of the transactions. Bulk supply refers to refinery production and the transport of refined products to terminals by pipelines, tankers or barges. Here we focus on the two largest volume categories of refined products--gasoline and No. 2 distillates.[13]

The Northeast accounted for approximately 16 and 21 percent of total U.S. demand for gasoline and distillates respectively in 2003, the year before Sunoco/El Paso. Northeast gasoline demand in 2003 averaged about 1,450 MBD, while distillate demand averaged 829 MBD.[14]

---

[12] FTC Valero/Kaneb complaint, at Paragraph 40.

[13] No. 2 distillates include No. 2 diesel fuel (used for automobiles, trucks and locomotives) and No. 2 fuel oil (used in residential and commercial heating). No.2 diesel fuel and No. 2 fuel oil are close production substitutes at the refinery level. Other, smaller volume refinery products include jet fuel, general aviation gasoline, kerosene, No. 1 and No. 4 distillates, and residual fuel oil. In our empirical analyses, we limit consideration to No.2 diesel fuel, the largest volume category in No. 2 distillates.

[14] The "Northeast" is defined as the six New England states, New York, Pennsylvania, New Jersey and Delaware. Gasoline and distillates consumption for this region are derived from U.S. Energy Information Administration ("EIA") data, available at:
http://tonto.eia.doe.gov/dnav/pet/pet_pnp_unc_dcu_nus_m.htm
http://tonto.eia.doe.gov/dnav/pet/pet_pnp_pct_dc_nus_pct_m.htm
http://www.eia.doe.gov/pub/oil_gas/petroleum/data_publications/petroleum_supply_annual/psa_volume1/historical/2003/pdf/table_36.pdf
http://www.eia.doe.gov/emeu/states/sep_use/notes/use_print2006.pdf

Seven refiners operated in the Northeast in 2003. Aside from the relatively small United Refining refinery in western Pennsylvania, all Northeast refineries were located on Atlantic Coast. (See Figure One.) Amerada Hess and ConocoPhillips each operated a refinery on the New Jersey side of the New York Harbor area. Six other refineries, accounting for 80 percent of regional capacity, were located on the lower Delaware River in the greater Philadelphia area. These refineries were controlled by five firms in 2003--Valero, ConocoPhillips, El Paso, Motiva, and Sunoco (with two refineries). Sunoco was the Northeast's leading refiner in 2003, with nearly 33 percent of the Northeast. (See Table One.) We estimate that Northeast refinery production accounted for roughly 70 percent of regional gasoline consumption and roughly 50 percent of distillates consumption in 2003.[15]

Imports from foreign refineries in 2003 accounted for approximately 22 and 30 percent of total Northeast consumption of gasoline and distillates respectively.[16] Most imports were landed in New England and in the New York Harbor area, along with a smaller quantity in the Philadelphia area. At least 25 firms imported significant quantities of gasoline and distillates into region in 2003, though about 70 percent of both gasoline and distillates imports in that year were accounted for by 5 firms. The top five gasoline importers in 2003 were Amerada Hess, BP, Citgo, Irving Oil, and Vitol.

---

http://www.eia.doe.gov/emeu/states/_seds.html
http://www.eia.doe.gov/oil_gas/petroleum/data_publications/company_level_imports/cli.html
http://tonto.eia.doe.gov/dnav/pet/pet_move_ptb_dc_R20-R10_mbbl_m.htm

[15] Individual refinery production data are not publically available. The Northeast production for gasoline and distillates were estimated using EIA data on 2003 Northeast refinery capacity, capacity utilization rates, and product percentage yields.
[16] Estimates are based on import data obtained from EIA.

Amerada Hess, Citgo, Global, Irving Oil, and Morgan Stanley were the top five distillates importers.[17]

Gasoline and distillate shipments from the Gulf, approximately 8 and 20 percent of Northeast consumption, completed Northeast supply.[18] Gulf product arrived in the Northeast largely on the Colonial Pipeline, with smaller quantities coming by tanker and barge.[19] The nation's largest refined products pipeline, Colonial connected refineries in Texas, Louisiana, Mississippi, and Alabama with terminals in the Southeast and Mid-Atlantic states before terminating in Linden, New Jersey. While quantity data on individual shippers into the Northeast from the Gulf are not publically available, many firms, including firms without Northeast refinery assets, sent or could have sent Gulf-refined gasoline and distillates into the Northeast.[20]

Within the Northeast, gasoline and distillates were distributed by pipeline, water, and (less frequently) by rail to about 180 product terminals (See Figure One.) Tank wagon trucks drew from terminal "racks" to supply retail service stations within a radius such as 50 to 75 miles.[21] The Laurel, Sun, and Mobil pipelines shipped Delaware River refinery product to terminals in western and south central Pennsylvania. Other segments of the Sun and Mobil pipeline systems moved products from the Delaware River to north central and north eastern Pennsylvania and to central and western New York. Another

---

[17] Company level imports are based on data from EIA.

[18] EIA does not directly report shipments of domestically refined product into the Northeast. Our estimates of domestically refined products shipped into the Northeast are based on EIA data on inter-PADD shipments, adjusting for apparent consumption of these shipments outside of the Northeast.

[19] Small quantities of gasoline and diesel may also have been barged into the Northeast from the Giant refinery in Yorktown, Virginia. Relatively small quantities of refined product may also have been barged on the Ohio River into the Pittsburgh area.

[20] Gulf refiners connected to Colonial, but having no Northeast refineries in 2003, included BP, ExxonMobil, and Chevron. Colonial reported over 80 shippers on its system in 2008, including Gulf area refiners, branded marketers not integrated into refining, and product traders. See Colonial Pipeline website at www.colpipe.com/ab_oc.asp.

[21] This distance depends on traffic congestion, fuel costs, demand density, and relative prices at other terminals.

segment of the Sun pipeline system and the Harbor Pipeline transported Delaware River refined product to the New York Harbor. Colonial could also pick up product from some Delaware River refiners for shipment north to the New York Harbor. Some Delaware River refineries also barged some product to the New York Harbor.

The Buckeye pipeline system was an important regional pipeline. From origins in the New York Harbor area, gasoline and distillates traveled west on Buckeye to north central New Jersey and into Pennsylvania. In Macungie, Pennsylvania, the Buckeye system branched north to supply terminals in northeast Pennsylvania and upstate New York. The southern branch served terminals in southern Pennsylvania and connected with the Laurel pipeline in Sinking Spring, Pennsylvania, approximately 65 miles northwest of Philadelphia. The Colonial pipeline connected with the Laurel pipeline at Booth, Pennsylvania. The Laurel pipeline was thus capable of shipping product originating from three sources: the Gulf, Delaware River refineries, and from the New York Harbor.

Some volumes of gasoline and distillates were redistributed from the New York Harbor area by barge or tanker to smaller ports such as Albany and New Haven. Though water redistribution from New York Harbor was generally to the north, shipments to terminals to the south, such as those on the Delaware River, may have sometimes occurred.

Domestically refined petroleum products arrived in New England primarily by tanker or barge, though limited quantities may have come from Canada and New York State by truck or rail. No pipeline connected New England to sources in the south, though several small pipelines linked New England ports to terminals in the interior. A Buckeye pipeline moved product from New Haven through Hartford to terminals near Springfield,

Massachusetts. An ExxonMobil pipeline carried refined product into central Massachusetts from East Providence, Rhode Island. Another ExxonMobil pipeline originated Portland, Maine and carried product north to Bangor.

New England relied more heavily on imports than the rest of the Northeast. In 2003, for example, foreign imports delivered to New England ports were about 57 and 60 percent of New England consumption of gasoline and distillates, respectively.[22]

To sum up, Northeast refinery capacity at the time of the mergers was primarily located on the Atlantic Coast, with most capacity concentrated on the lower Delaware River. Though accounting for most of regional supply, regional refiners had insufficient capacity to meet regional demand. Shipments from the Gulf on Colonial and foreign imports, which were primarily landed in New England and the New York Harbor area, rounded out Northeast supply. Pipelines, tankers and barges, and to a lesser extent rail, were used to move bulk quantities of refined product within the Northeast, though the logistics in supplying particular terminals with the region varied.

III. The FTC's Antitrust Concerns.

The FTC's closing statement on the Sunoco/El Paso investigation identified the two "most plausible concerns of anticompetitive harm": 1) an increase gasoline prices in the Philadelphia area, and 2) an increase in gasoline prices along locations served by the Laurel pipeline, an area in central Pennsylvania which the FTC referred to as the "Laurel Corridor."[23] The FTC concluded that an anticompetitive, Philadelphia-area price increase by local refiners would be defeated by diversion of Colonial shipments from

---

[22] Based on EIA import and consumption data. See footnote 15.

[23] Statement of the Commission in the Matter of Sunoco Inc./Coastal Eagle Point Company, File No. 0310139, December 29, 2003, (hereafter referred to as the FTC statement on Sunoco/El Paso) available at: http://www.ftc.gov/os/caselist/0310139/031229stmt0310139.pdf. The FTC does not usually issue statements when investigations are closed. No statement was issued in Valero/Premcor.

delivery in the New York Harbor area to Philadelphia area terminals. The FTC concluded

that, because the New York area had ample supply from alternative sources, a diversion

of Colonial supply into Philadelphia would not cause price increases in the New York

area. Increased waterborne shipments into the Philadelphia area, particularly from the

U.S. Virgin Islands, would also keep prices at a competitive level, according to the FTC

statement.[24] As for the Laurel Corridor, the FTC analyzed "expressions of concern" that

Delaware River refiners could increase post-merger prices in the Corridor because other

potential sources, Buckeye and Colonial, were full during the summer, were not

"economically viable" relative to Delaware River refiners, or were otherwise

"constrained by certain logistical impediments." Upon further investigation the FTC

concluded that any post merger attempt by Delaware River refiners to raise Corridor

gasoline prices anti-competitively would not succeed because of increased shipments into

the Corridor via the Colonial Pipeline.

The FTC's complaint in Valero/Kaneb provides information about the agency's

competitive concerns in this region and therefore provides additional insight regarding

Sunoco/El Paso and Valero/Premcor. Kaneb operated three terminals, two in

Philadelphia and one in nearby Paulsboro, New Jersey. Among other things, the FTC

alleged that the Valero/Kaneb merger would have adverse competitive consequences at

the bulk supply level.[25] The Kaneb terminals were connected to Colonial. One of

---

[24] The Hovensa refinery, located in the U.S. Virgin Islands, had an operable capacity in 2003 of 470 MBD, making it one of the largest refineries in the world. Hovensa was a joint venture of Amerada Hess and Petroleos de Venezuela, S.A. Amerada Hess operated a product terminal in the Philadelphia area capable of receiving deep water cargoes.

[25] The FTC also alleged that Valero and Kaneb were direct horizontal competitors in providing terminalling services for refined products in the greater Philadelphia area. Valero operated a product terminal in Paulsboro, NJ, which was supplied by its adjacent refinery. This competitive concern was independent of the transaction's competitive implications at the bulk supply level to the extent that terminals in the area only provided terminalling services for locally refined product.

Kaneb's Philadelphia terminals also could receive product by barge, and the Paulsboro terminal was capable of receiving bulk shipments by deepwater tankers. The FTC alleged that shippers using the Kaneb terminals imposed a competitive constraint upon Philadelphia area prices such that area prices would be generally limited by "either Gulf prices plus pipeline tariff or New York Harbor prices adjusted by water-borne transportation costs..."[26] Absent this constraint, the FTC alleged that a combined Valero/Kaneb could effectively coordinate with other Delaware River refiners or terminal operators (presumably including Amerada Hess, which also operated an terminal in the area) to raise Philadelphia area prices. As earlier discussed, the FTC entered into a consent agreement under which Kaneb's Philadelphia area terminals would be divested.

The FTC's analysis in Valero/Kaneb presumed that Valero/Premcor had been consummated, indicating that the agency believed that the constraint imposed by Colonial and water shipments into the Philadelphia area were sufficient to maintain the level of pre-merger competition among the three remaining local refiners.[27]

As stated in the FTC complaint in Valero/Kaneb, the leading competitive concern was collusion among Delaware River refiners to raise gasoline prices in selected locations in the Northeast. The agency apparently concluded that price increases in other parts of the Northeast (or for the Northeast overall), or for refined products other than gasoline, were less plausible.[28] Implicit in the FTC's enforcement rationale is the assumption that successful collusion among a broader group of firms that also included

---

[26] FTC Complaint in Valero/Kaneb, at Paragraph 38.

[27] *Ibid.*, at Paragraph 36.

[28] The FTC statement on Sunoco/El Paso noted that it did not discuss all the potential markets for which the FTC conducted an investigation, only those that raised the most plausible anticompetitive concerns.

Colonial shippers and importers was unlikely, and absent any significant constraints on these alternatives, the mergers would not likely result in anticompetitive price increases.

We test the FTC's enforcement rationale in these mergers by evaluating the following questions:

1) Did gasoline or diesel prices in the greater Philadelphia area rise after the Sunoco/El Paso and Valero/Premcor mergers compared to other prices elsewhere in the Northeast[29]?

2) Did gasoline or diesel prices along the Laurel Corridor go up after the mergers compared to prices at other Northeast locations?

3) Did gasoline or diesel prices in the Northeast increase post-merger relative to prices outside the Northeast?

IV. Methodology and Data.

The most common empirical strategy to identify merger price effects is some form of a difference-in-difference ("DID") estimator.[30] Prices in areas potentially affected by a merger ("treatment" areas) are compared to prices in unaffected areas ("control" areas) that have similar demand and cost changes as those in treatment areas. Differences in the pre- and post-merger price difference between treatments and controls ideally identify merger effects because common cost and demand shocks are netted out.

A. Baseline Model

---

[29] We do not separately test for effects from Valero/Kaneb. First, Valero/Premcor and Valero/Kaneb were consummated nearly at the same time, thus making it difficult to segregate any effects from the two mergers. Second, because the FTC entered into a consent requiring divestiture of Philadelphia area terminals in Valero/Kaneb, anticompetitive effects might not be expected. However, if for some reason the divestiture did not maintain the competitive status quo, any price effects from Valero/Kaneb would be limited to the Philadelphia area, and in our analyses these would be attributed to Valero/Premcor.

[30] For examples of this approach involving non-petroleum industry mergers, see Barton and Sherman (1984), Kim and Singal (1993), and Vita and Sacher (2001). See Hastings (2004), Hastings and Gilbert (2005), Taylor and Hosken (2007) and Simpson and Taylor (2008) for examples involving petroleum industry mergers.

We assume that the price of gasoline (or diesel fuel) in an affected area (p_{At}) can be explained by changes induced by Sunoco/El Paso and Valero/Premcor, seasonal effects (proxied by month dummies, D_{mt}), and time-specific supply and demand shocks (γ_t) as described by equation (1) below. We make the usual assumption that the transactions are exogenous.

$$(1) \; p_{At} = \alpha_0 + \alpha_1 \text{Sunoco}(1)_t + \alpha_2 \text{Valero}(2)_t + \sum_{m=1}^{11} \beta_m D_{mt} + \gamma_t + \varepsilon_{At}$$

The prices in the control areas (p_{Ct}) are explained by a similar relationship described by equation (2) below:

$$(2) \; p_{Ct} = \theta_0 + \sum_{m=1}^{11} \lambda_m D_{mt} + \gamma_t + \varepsilon_{Ct}$$

To estimate the price effects of the transactions, we take the difference of equations (1) and (2) and estimate equation (3) below, which eliminates common, time-specific shocks (γ_t).

$$(3) \; p_{At} - p_{Ct} = (\alpha_0 - \theta_0) + \alpha_1 \text{Sunoco}(1)_t + \alpha_2 \text{Valero}(2)_t + \sum_{m=1}^{11} (\beta_m - \lambda_m) D_{mt} + (\varepsilon_{At} - \varepsilon_{Ct})$$

We allow for differing price levels in affected and control areas $(\alpha_0 \neq \theta_0)$ and for systematic differences in seasonal pricing $(\beta_m \neq \lambda_m)$.[31] Because the error term of equation (3) is autoregressive, we employ an AR(1) correction.[32] Estimates of α_1 and α_2 may be either positive or negative depending on whether the merger was anticompetitive or, on the other hand, led to lower prices due to merger-related efficiencies.

B. *Control Areas*

---

[31] For example, there are regional differences in seasonal changes in gasoline prices, driven by such factors as pipeline constraints and summer/winter differences in formulations.
[32] We use the Prais-Winsten correction for autocorrelation.

Identification of merger price effects requires common, time-specific supply and demand shocks (γ_t) in treatment and control areas. If treatment and control areas instead experience persistently different demand or cost shocks, disentangling any merger price effects from any demand or cost changes is impossible.[33] Most of the variability in gasoline and diesel prices is attributable to changes in crude oil prices. Because refined product prices everywhere are similarly sensitive to crude oil price changes, choice of control area is not very critical to account for shocks related to crude. The bigger challenge is designating control areas that share other important cost and demand shocks affecting refining and bulk transport, including outages, capacity constraints, and changes in transportation charges or in the refiners' and marketers' opportunity costs in geographically allocating product. Areas close to a treatment area would more likely share these demand and costs shocks, but relatively close areas may be less than ideal controls because their prices might be impacted by a merger-related price change in a treatment area due to geographic arbitrage.

Acknowledging these tradeoffs, we use multiple, alternative control areas of varying proximity for each affected area to assess the robustness of our results. We have generally designated controls far away enough from affected areas such that price differences are unlikely to be arbitraged by consumers at the retail level or by distributors at wholesale who might divert tank trucks from terminals in a treatment area to terminals in a control area.

---

[33] Not only should the areas experience the same shocks, but the pass through of the shocks from one price level to the next needs to be the same. See Simpson and Schmidt (2008).

We also pair treatment and control areas with the same gasoline specification because time-specific shocks across different gasoline specifications may vary.[34] At the time of the mergers, federal environmental regulations required that the entire states of Delaware, New Jersey, Connecticut, Rhode Island, and Massachusetts and parts of New York, Pennsylvania and New Hampshire use "reformulated" gasoline ("RFG"), a more expensive, but less polluting specification than conventional gasoline. Conventional gasoline was sold in all other parts of the Northeast at the time of the mergers, though the greater Pittsburgh area and southeast Maine required a variant of conventional gasoline (low Reid Vapor Pressure gasoline, sometimes referred to as "7.8 grade") for summer time use.

Arbitrage between our designated control and treatment areas at the bulk supply level might still occur. For example, in response to a merger-related price in the Northeast, Gulf refiners and other marketers might divert pipeline shipments to the Northeast from control areas that we have designated in the Southeast and Midwest. However, we assume, that for any plausible merger-related output reduction in any affected Northeast location, any resulting arbitrage at the bulk supply level would be spread over such a broad area that any price effect in specific control areas is *de minimus*.

1. Northeast Controls for Philadelphia. We select Boston and Newark as controls to test whether Philadelphia post-merger prices changed relative to other Northeast locations. These three areas consistently used the same gasoline specification

---

[34] For example, refiners might differ, at least in the short run, in their capabilities to produce different gasoline fuel specifications, raising the possibility of different supply shocks across gasoline specifications should there be refinery outages.

during the period--RFG North.[35] We also used Boston and Newark as controls in our diesel analysis, although diesel regulatory requirements were not similarly geographically differentiated. Though close enough to Philadelphia to raise some questions about its independence as a control due to tank truck arbitrage, Newark is of interest due to the FTC's conclusion that New York Harbor area prices would not increase after the Sun/El Paso merger because of competition from imports and the two New York Harbor refiners.

2. Northeast Controls for the Laurel Corridor. We measure prices in the affected Laurel Corridor at two points—Harrisburg and Pittsburgh. Harrisburg and most of the rest of Pennsylvania used standard conventional gasoline. Albany, NY and Bangor, Maine used the same fuel specification as Harrisburg, and we utilize those areas as Northeast controls for Harrisburg. Pittsburgh also used conventional gasoline, but with a special 7.8 RVP mandate during the summer. Having the same RVP restrictions during the period, Portland, Maine was our only Northeast control choice for Pittsburgh. We use the same control areas for Harrisburg and Pittsburgh in our diesel analysis. Interest in 7.8 RVP is motivated by the possibility that market power might be more easily exercised in relatively low volume, "boutique fuels."

3. Outside of Northeast Controls for Northeast Prices. To test whether Northeast prices rose relative to prices outside the Northeast, we again group treatments and controls by gasoline specification. We compare gasoline prices in Philadelphia, Newark, and Boston to RFG prices in other parts of the U.S. However, aside from Louisville, KY which also requires RFG North and which we use as a control, other parts

---

[35] We ruled out controls in New York and Connecticut because, although these areas also used RFG North, they switched to ethanol from MTBE as a gasoline oxygenate by December 31, 2003, a date nearly coincident with Sunoco/El Paso transaction. Other parts of the Northeast using RFG North did not switch to ethanol as an oxygenate until summer of 2006.

of the U.S. requiring RFG used the slightly different RFG South formulation.[36] Because RFG South and RFG North appeared to be close substitutes in production (closer than conventional gasoline or diesel), we use the RFG South locations of Fairfax, Virginia and Houston, Texas as controls for Northeast RFG prices, though we recognize that RFG North/South differences somewhat weaken these controls. In conventional gasoline we pair the treatment area, Harrisburg, with the controls areas of Charlottesville, Virginia, Roanoke, Virginia, and Lexington, Kentucky. Outside-of-Northeast controls for Pittsburgh and its 7.8 RVP gasoline are Detroit, Michigan and New Orleans, Louisiana. The same out-of-Northeast controls are used in the diesel analysis.

C. *Merger Windows*

Our analysis requires a pre-merger period sufficiently long to estimate pre-merger price relationships between treatment and control areas, and a post-merger period sufficiently long to allow firms to take advantage of any merger-related market power or efficiencies. The post-merger period cannot be so long, however, as to pick up non-merger related changes in market conditions that might affect relative prices in treatment and control areas. In our baseline estimates we use a two year window prior to January 13, 2004, the day Sunoco/El Paso was consummated.

We assume that merger price effects may occur immediately upon consummation. The Sunoco/El Paso post-merger window in our baseline estimates is twenty months, ending on September 1, 2005, the consummation day for Valero/Premcor. The possible merger effect for Valero/Premcor is measured from September 1, 2005 until the end of

---

[36] Northern and Southern RFG have different volatile organic compound ("VOC") emission standards during the summertime. See U.S. Environmental Protection Agency, "Study of Unique Fuel Blends ("Boutique Fuels"), Effects on Fuel Supply and Distribution and Potential Improvements," October 2001, available at http://www.epa.gov/otaq/regs/fuels/p01004.pdf at 85-86.

the data in December 2007. Note that if there were a Sunoco/El Paso effect that did not show up until after consummation of Valero/Premcor, that effect would be reflected in the Valero/Premcor merger estimate. Recognizing that merger effects might not occur immediately, we also test for effects with windows opening three and six months after consummation as a robustness check.[37]

D. *Measurement of Price*

The mergers did not increase control of competing retail outlets and directly implicated competition only at the bulk supply level. As such, the transactions' primary effect should be upon wholesale prices. However, suitably disaggregated available data on wholesale gasoline and diesel prices are limited to wholesale rack prices: those prices paid by distributors at product terminals. Other wholesale prices, for which public data are more limited or totally unavailable, include bulk spot prices (arm's length, individual sales of large quantities of gasoline or diesel), refinery gate price (FOB prices for specified volumes or range of volumes set under negotiated contracts of various durations), dealer tank wagon prices (prices set by refiners and other marketers for delivery of gasoline and diesel to individual service stations), and internal transfer prices (for refiners and marketers who own and operate their own service stations). If merger-

---

[37] Some analysts have used a post-merger window beginning at a transaction's announcement date (GAO (2009)). We think a post-merger window beginning at the announcement date is unrealistic because of the uncertainty that the transaction will be completed due to either antitrust challenge or purely business related reasons and because of significant penalties should antitrust enforcers uncover any attempts to jointly control the merging firms prior to consummation. Such "gun jumping" may be detected during prospective review by antitrust authorities, and merging firms may be liable for penalties even if the merger itself goes unchallenged.

Effects beginning sometime after consummation might be expected for several reasons. Refinery output slates are largely determined up to several months in advance as refiners seek to lock in crude oil and other input purchases. Pipeline nominations are also made on an advance basis, and some contracts with bulk purchasers may limit refiners' ability to adjust output immediately. Moreover, if post-merger collusion were a competitive concern, some time might pass before coordinating rivals reached a consensus on prices. Finally, even a longer period of time might be required for firms to capture any merger-related efficiency gains.

related price effects vary across these different wholesale prices, basing the analysis just on rack prices may yield misleading results. The net effect of any changes across all wholesale prices should be reflected in retail prices, however. Consequently, we test for both retail and wholesale rack price effects.

Our price data comes from the Oil Price Information Service ("OPIS"). OPIS collects data on retail and wholesale rack prices for numerous areas. Rack prices consist of the daily average price for branded and unbranded gasoline and diesel at terminal locations based on OPIS' survey of terminal operators. OPIS' retail data is derived from service stations that accept corporate fleet cards. We use the OPIS constructed average retail price for specific OPIS-designated areas. OPIS calculates this price as the average price over all retail outlets in an area that report on a give day, e.g., all stations in the Pennsylvania portion of the Philadelphia area. While OPIS retail price data are among the best available, they do not represent a random sample of retail outlets, and not every outlet may report on every day.

OPIS' retail price includes taxes, but we remove all federal, state and local taxes. We aggregate the daily data to the weekly level, in part to mitigate any changes in sample composition that might arise from day to day changes in the number and identity of reporting retail outlets. For comparability, we also aggregate daily wholesale prices to the weekly level by taking the weekly average of the daily average branded and (separately) the unbranded low wholesale OPIS prices.

Most gasoline sold in the U.S.--approximately 80 percent in 2002--is regular octane gasoline, and thus we focus on the regular gasoline in this study. There is only one grade of diesel fuel, although its specification has changed over time.[38]

E. *The 2005 Hurricanes*

The supply disruptions from two major Gulf area hurricanes of 2005 would likely confound our merger effect estimates. Hurricane Katrina came ashore on August 29, 2005 and Hurricane Rita struck on September 24, 2005. These storms devastated Gulf Coast refinery and pipeline infrastructure and resulted in large price spikes for refined products throughout the U.S, as well as temporarily widening price differences among geographic areas. To control for these hurricane effects, we include a week-specific dummy, and in our baseline estimates we designate the weeks from September 1 through the end of November 2005 as hurricane-affected. While we report results without the hurricane control below, we believe that estimates controlling for the hurricanes to be more probative.[39] We also later report on robustness checks in varying the duration of the hurricane affected period.

V. Results

A. *Descriptive Statistics*

Table Two presents descriptive statistics on gasoline and diesel prices grouped by treatment area and measure of price. Reading right to left, the first row of each grouping shows the number of weekly observations, the treatment area's mean price, standard

---

[38] A formulation change from low-sulfur to ultra-low sulfur diesel was mandated by the U.S. Environmental Protection Agency ("EPA") in 2006 for on-highway diesel, and in 2007 for off-road diesel. Most states completed this change by the end of summer 2006. See EPA's Direct Final Rule, available at, http://www.epa.gov/otaq/regs/fuels/diesel/420f06033.htm.

[39] See U.S. Federal Trade Commission, *Investigation of Gasoline Price Manipulation and Post-Katrina Gasoline Price Increases*, (Spring 2006) (hereafter referred to as "FTC Katrina Report"), available at http://www.ftc.gov/reports/060518PublicGasolinePricesInvestigationReportFinal.pdf.

deviation, and its minimum and maximum prices. Rows below in each grouping report the mean price difference between the treatment and control areas, this difference's standard deviation, and the minimum and maximum differences. For example, Philadelphia's mean (less tax) retail gasoline price was $1.62 per gallon over the period, ranging from a low of 60 cents per gallon to a high of $2.84. The mean difference in gasoline between Philadelphia and the control areas was no more than about 5 cpg, although there was considerable variation in the minimum and maximum differences across controls. Standard deviation of treatment/control price differences, as well as minimum and maximum treatment/control differences varied across pairings. For example, the standard deviations of Philadelphia-Boston and Philadelphia-Newark price differences (3.1 and 3.5 respectively) were smaller than those for the other Philadelphia controls of Fairfax (5.1), Louisville (11.2) and Houston (7.4).

Table Two also shows that both branded and unbranded rack prices have smaller average treatment/control differences as compared to retail. Standard deviations of the differences in rack prices are also generally lower than those at retail, which is not surprising because the rack prices do not include variation in the retail markup. Unbranded pricing differences are more volatile than the branded pricing differences as expected because unbranded prices typically react more to supply disruptions or shortages than branded prices.[40] The average price for the different gasoline specifications are in the order expected based on differences in refinery costs: the average branded conventional price (in Harrisburg) is about two cents per gallon less than boutique, low RVP conventional (Pittsburgh) and about five cents less than the average

---

[40] See Bulow *et al.* (2003) for a discussion of relationship between branded and unbranded prices and their reactions to supply disruptions.

price of RFG (Philadelphia).[41] Diesel price relationships and standard deviations among treatment/control pairs are generally similar to those for gasoline, adjusting for the higher average price of diesel compared to gasoline.

B. *Baseline Results.*

Table Three presents estimated price effects for the Sunoco/El Paso and Valero/Premcor based on control areas within the Northeast. Table Four shows estimated effects in treatment areas based on controls outside the Northeast. Rows 2, 4, and 6 of the tables show estimated effects on retail, branded rack, and unbranded rack prices, taking into account the hurricane period, while estimates in rows 1, 3, and 5 are without the hurricane control. It is obvious that the estimates can be very sensitive to controlling for the hurricane period. As noted above, we believe the estimates controlling for the hurricanes are more probative of the mergers' possible effects, and we restrict our discussion to those results.

Turning first to Sunoco/El Paso, we find no statistically significant, positive estimate for retail prices for any treatment area, although statistically significant negative estimates are found in two instances (Pittsburgh/Portland in diesel, and Harrisburg/Roanoke in gasoline). Sunoco/El Paso was not associated with any significant price effect in branded rack prices in any treatment/control pairing. Estimates for unbranded rack were not generally significant, but there were several exceptions: significant positive estimates in gasoline for the Philadelphia/Newark pairing and in diesel for Harrisburg/Bangor, Pittsburgh/Portland and Harrisburg/Charlottesville; a small but statistically significant effect was found in diesel for Harrisburg/Albany. On net,

---

[41] Lidderdale and Bohn (1999) estimate that phase II RFG would cost four to five cents more per gallon than conventional gasoline.

these baseline results indicate that Sunoco/El Paso had little or no important anticompetitive effect. They suggest no positive impact on either retail or branded rack prices, a finding that was robust across all control areas. Estimates for unbranded prices were more mixed, but even here most controls suggested no significant effect.

Results for Valero/Premcor are somewhat more complex.[42] The results suggested no positive price impact upon Laurel Corridor prices, and, as Table Four shows, in a number of instances Harrisburg and Pittsburgh prices appear to have significantly fallen relative to control areas outside of the Northeast--nearly all relative price declines were in diesel. Any anticompetitive effects of Valero/Premcor appear limited to Philadelphia prices, although the evidence for a Philadelphia effect is mixed. Table Three shows merger-related, retail price increases of about 3 cents per gallon in Philadelphia relative to Boston in both gasoline and diesel; a similar increase in Philadelphia diesel (but not in gasoline) relative to Newark was also found. Retail estimates using the outside of Northeast controls of Fairfax, Louisville and Houston were not significant, however. At rack, the Valero/Premcor transaction was estimated to have a significant positive effect upon Philadelphia branded diesel prices relative to Newark, while the estimated effect on Philadelphia unbranded diesel prices relative to Newark was significantly negative (and surprisingly large). Unbranded Philadelphia rack gasoline prices appeared to have risen relative to Newark post merger. Estimated Valero/Premcor rack price effects for all other treatment/control pairs were either insignificant or significant and negative.

C. *Discussion and Robustness Checks*

---

[42] As noted previously, these estimates may also reflect any effects from Sunoco/El Paso that occurred after consummation of Valero/Premcor.

Identification of merger effects depends critically on choice of control areas. Because no control is ideal, evaluating post merger effects using multiple alternatives is an important robustness check. Our baseline results provided 168 merger effects estimates. The vast majority (138 out of 168) was statistically insignificant, while 21 estimates were significant but negative. Only 9 estimates suggested a significant positive price increase, and just 3 of these were at retail. Thus the preponderance of evidence, based on a count of significant estimates among treatment/control pairings, suggest that the mergers were, at worst, competitively neutral.

Differences in the price relationship of the treatment and control areas caution against drawing inferences only from a count of significant results across control alternatives. Prices in treatment areas are more closely related to those in controls within the Northeast compared to outside controls. This pattern is consistent for the three price levels for both gasoline and diesel. As Table Two indicates, for example, the standard deviation of Philadelphia/Boston RFG price difference (3.1 at retail) is noticeably smaller than the standard deviation in the Philadelphia/Fairfax price difference (5.1), despite Fairfax and Philadelphia being closer geographically. Similarly, in conventional gasoline, Harrisburg retail prices are more closely related to those in Bangor, Maine than in Roanoke, Virginia (standard deviation in the difference from Bangor of 2.8, compared to 5.1 for the difference from Roanoke). We interpret these differences in the tightness of price relationships as indicating that, although both Gulf and foreign imports are shipped into the regions, imports are generally the more important of the two in determining Northeast prices. This finding is contrary to the FTC's assumption in evaluating these transactions, which as noted above, viewed Gulf product as the chief potential constraint of any anticompetitive behavior by Northeast refiners.

These considerations also imply that results with Northeast controls should be given greater weight in assessing the mergers' competitive effects. While the Northeast controls may not be ideal because their prices may be more likely affected by the transactions compared to out of the Northeast control area, the impact of any such effect should be to bias *against* finding significant, merger related effects. But Table Three reports a number of significant effects, including several positive estimates of between one and three cents per gallon for Philadelphia area prices. Because the results with Northeast controls may be more telling about the mergers' competitive effects, we focus our robustness checks on the regressions in Table Three.

1. Robustness and Timing Assumptions. We first apply robustness checks that vary our baseline timing assumptions on 1) the length of the pre-merger period, 2) the time when the mergers were assumed to affect prices, and 3) the duration of the hurricane-affected period. That the baseline merger estimates might be sensitive to timing assumptions is illustrated by Figure Two. Figure Two depicts the timing assumptions in the baseline model along with, as an example, retail price differences between Philadelphia and three control areas-Newark, Boston and Fairfax. While there is volatility in all price difference series, the tighter price relationships between Philadelphia and Newark and Philadelphia and Boston compared to Philadelphia versus Fairfax is evident. The extreme volatility in price differences during the hurricane period is particularly notable. Differences in other treatment and control prices exhibit similar volatility over the period of the data.

Table Five reports the results of varying our timing assumptions for Sunoco/El Paso, while Table Six reports comparable findings for Valero/Premcor. Column Two identifies significant merger effect estimates in the baseline model for all treatment/control/price combinations, while columns to the right indicate whether significant merger effect results were obtained by altering timing assumptions. The "pre-merger period" column reports results when the baseline model is re-estimated by dropping 2002, thus using a pre-merger period of just 2003. Columns 4 and 5 report results when the baseline model is re-estimated by delaying the effective date for possible price effects by three and six months respectively after the mergers' actual consummation. Columns 6 and 7 show results in changing assumptions about the duration of the hurricane-affected period. The re-estimation in Column 6 assumes the a hurricane period beginning two weeks earlier than in the baseline to capture any anticipatory price changes; the estimates in Column 7 assumed that the hurricane period lasted an additional three months beyond the baseline assumption so as to capture any lingering effects of the storms.

Tables Five and Six show that the results are largely robust to these changes in timing assumptions. Results that were statistically insignificant under the baseline generally remained insignificant under the timing assumption alternatives. There were several exceptions to this (e.g. Philadelphia/Newark gasoline and diesel retail in Sunoco/El Paso) but none of these results were robust across the timing alternatives. With some exceptions (e.g. Philadelphia/Newark diesel retail in Valero/Premcor), estimates that were significant in the baseline generally remained significant (and with the same sign) under the timing assumption alternatives.

27

One robust significantly positive result in both transactions is the positive effect in

unbranded RFG gasoline in the Philadelphia/Newark comparison. While this positive

result was not observed in the other RFG comparison (Philadelphia/Boston), we note that

the only other study to examine any of these transactions, GAO (2009), found that the

Valero/Premcor merger was also associated with a statistically significant price increase

in unbranded rack gasoline of 1.1 cents per gallon, but –similar to our results--no

statistically significant effect for branded rack.[43]

Significantly positive retail price effects in the Philadelphia/Boston comparison

in Valero/Premcor were also relatively robust. We found this result surprising in the

absence of positive rack price effects, although, as discussed above, such an outcome

might occur because not all wholesale prices are observed. As can be seen in Figure Two,

the retail price of gasoline was increasing in Philadelphia relative to Boston in the pre-

merger period. This trend is even more pronounced in the retail diesel data, but the causes

for these trends are unclear. To further examine whether retail prices changed in

Philadelphia, we re-ran the regressions with three other New England areas--Barnstable,

Massachusetts, Portsmouth, New Hampshire and Providence, Rhode Island--as controls.

We did not detect any significant increase retail price of gasoline or diesel in Philadelphia

after Valero/ Premcor relative to any of these three alternative control areas.[44]

2. Year Effects

We also checked the robustness of the baseline results by comparing one full year

before the transactions, 2003, to one year after each transaction, 2004 for Sunoco/El Paso

---

[43] GAO (2009) took into account rack prices at more locations than in our study, but did not control for
supply disruptions from the 2005 hurricanes. As Tables Three and Four show, the hurricanes mattered,
especially for the unbranded rack estimates.
[44] We do not have rack prices for these three additional New England areas.

and 2006 for Valero/Premcor. This regression had the same dependent variable as equation (3) and has year dummy variables instead of merger specific dummy variables.

$$(4)\ p_{At}\text{-}p_{Ct} = (\alpha_0 - \theta_0) + \alpha_1 2004_t + \alpha_2 2005_t + \alpha_3 2006_t + \alpha_4 2007_t + \sum_{m=1}^{11} (\beta_m - \lambda_m) D_{mt} + (\varepsilon_{At} - \varepsilon_{Ct})$$

The regression was estimated using data from 2003-2007. Sunoco/El Paso was consummated in the first weeks of 2004. Valero/Premcor was consummated right after Hurricane Katrina in 2005. The coefficient α_1 shows the year effect for 2004 relative to 2003, one year before and after Sunoco/El Paso. The coefficient α_3 shows the year effect for 2006 relative to 2003, one year before either transaction, relative to the first full calendar year after Valero/Premcor. While this estimation uses a different amount of data in both the pre and post period than in the baseline results, it uses symmetric amounts of data in the pre and post periods for complete calendar years.

The results of this estimation for the Northeast are given in Table Seven. With respect to retail prices, the only year coefficient estimate for either 2004 or 2006 for the Northeast treatment/ control pairs that is statistically significant is a decrease in the price of diesel in Harrisburg relative to Bangor. The estimated relationship for retail diesel fuel in Philadelphia relative to Boston for 2006, Valero/Premcor, is not significantly different from zero but it is positive and significant for 2005 and 2007.

Almost all of the estimated rack relationships in the baseline results are qualitatively unchanged from this robustness check with the exception of the positive and significant effect estimated for the unbranded rack in Harrisburg relative to Bangor for Sunoco/El Paso, which is no longer significantly different from zero.

3. Summary of Merger Effect Point Estimates

Tables Eight and Nine summarize our price change point estimates across both mergers. Table Eight separately summarizes the baseline model results for the retail, branded rack and unbranded regressions across all treatment/control pairs and across both gasoline and diesel. The table's last two columns reflect baseline model point estimates across all three price levels and across both fuels for inside and outside Northeast control regressions separately. As Table Eight shows, a minority of point estimates for the baseline model was positive in all instances. In all instances very few of these positive point estimates were strongly statistically significant. To the extent there were significantly positive estimates, these occurred only in regressions using inside Northeast controls and were most frequent in the unbranded regressions. As for the frequency distribution of point estimates, estimates between -1 and +1 cpg were most common followed by estimates ranging between -1 and -5 cpg. Big estimates—more than plus or minus 5 cpg, were very few in number.

Table Nine compares summaries of price change point estimates for our six robustness checks across all three price levels and across both fuels. (As may be recalled, our robustness check regressions are limited to inside Northeast controls.) Percentages of positive results, and percentages of strongly significant positive results, varied modestly across robustness checks. As with the baseline model, point estimates under all robustness checks were most frequently in the -1 to +1 cpg range, with the -1 to -5 cpg range having the next highest number of estimates. Large positive or negative estimates were relatively infrequent and were mostly limited to large negatives associated with the year effects regressions.[45]

D. *Gulf and Northeast Spot Price Comparisons.*

We found little evidence that Northeast prices increased relative to prices outside the Northeast after the mergers controlling for the Hurricanes. If anything, relative prices in the Northeast, particularly for diesel, went down.[46] As an additional check on whether the post-merger prices in the Northeast changed relative to prices outside the Northeast, we examined New York Harbor and Gulf spot prices for both gasoline and diesel. Spot prices involve individual transactions on the order of thousands of barrels and occur at transfer points such as refineries, ports, and pipeline junctures. Spot prices are important determinants of wholesale rack prices, and rack prices are highly responsive to changes in spot prices.[47]

Table Ten presents DID merger price effect estimates for New York Harbor spot prices (treatment) relative to Gulf spots (control) for conventional and RFG/RBOB gasoline and for diesel. Estimates without and the baseline hurricane control period are shown, along with estimates generated if a number of additional price spikes are excluded in addition to the hurricane period[48] As the Table shows, neither merger was associated

---

[45] Considering the estimates for the two mergers separately yield broadly similar conclusions. One notable exception, however, is that the modal price change estimates in Sunoco/El Paso regressions were between -1 and +1 cpg, while the modal range in Valero/Premcor was between -1 and -5 cpg.

[46] One exception was a marginally significant 1.4 cpg increase in Harrisburg's unbranded diesel rack prices relative to Charlottesville in Sunoco/El Paso.

[47] FTC Katrina Report at 99.

[48] We identified seven periods during the six years of data of positive and negative price spikes, each lasting between 2 and 5 weeks. The regressions reported in the bottom box of Table Ten remove these seven periods. There were two time periods before the first transaction, late February through early March 2003 and the middle of August thorough the middle of September 2003 -- where the price differences between New York Harbor and the Gulf were large by historical standards, more than 10 cents per gallon. Removing just these two periods (eight observations) from the pre-transaction period changes the significance of the reformulated gasoline estimated price change but does not appreciably change the estimated effect and significance of the diesel price change.

with a significant change in New York Harbor spot conventional gasoline prices relative to Gulf spot, and all point estimates were close to zero. Sunoco/El Paso estimates for New York Harbor RFG/RBOB also are all insignificant with point estimates close to zero. RFG/RBOB and diesel prices may have declined relative to the Gulf after Valero/Premcor. The Valero/Premcor RFG/RBOB estimates, however, are not robust to removing a small number of pre-and post-transaction observations. The estimated change in the relative spot price of diesel fuel associated with Valero/Premcor--approximately negative two to three cpg--is not appreciably sensitive to controlling for the hurricanes or for other pre-and post-merger spikes.

VI. Conclusions

Sunoco's 2004 acquisition of El Paso's New Jersey refinery and Valero's 2005 acquisition of Premcor's Delaware refinery significantly consolidated refinery control in the U.S. Northeast. The FTC investigated these transactions but challenged neither, in large part because the agency perceived that shipments from the Gulf of Mexico would constrain any anticompetitive behavior by Northeast refiners.

Examining prices for gasoline and diesel at both the retail and wholesale levels, our findings across multiple treatment and control areas generally suggest that the transactions were at worst competitively neutral. A few results indicated that some unbranded rack prices may have increased relative to other Northeast prices after the mergers. However, this outcome was not robust across controls or assumptions, and these unbranded price increases were not accompanied by branded rack or retail price increases. Northeast prices did not generally increase relative to prices outside the Northeast after the transactions. Differences in the closeness of the price relationships

32

between various treatment and control areas suggest that, contrary to the FTC's view at the time of the transactions, imports were generally more important in determining Northeast prices—and in constraining any anticompetitive behavior by regional refiners—than shipments from the Gulf. Additional analysis of the relative importance of Gulf and import supply may be warranted, however.

Many factors affect gasoline and diesel prices. We suspect that the impact from petroleum mergers upon prices is likely to be small relative to many of these other factors. The success of the DID approach in disentangling merger impacts from other factors affecting prices depends critically upon selection of control areas. No single control is likely to be ideal. Identifying good controls with time-specific cost and supply shocks common to treatment areas may be particularly challenging for refinery mergers because such mergers may affect prices over a broad geographic area in which supply and demand conditions may vary. Evaluating possible merger effects with multiple, reasonably plausible, controls is clearly warranted under these circumstances. Our analysis also points out the necessity of controlling for supply shocks, such as hurricanes, which can differentially affect treatment and control areas.

www.ingramcontent.com/pod-product-compliance
Lightning Source LLC
Chambersburg PA
CBHW081245170526
45165CB00009B/3202